CHICKENS!

A Homeschool Curriculum on The History, Care, and Raising of Chickens That Supports K-3 Educational Standards with Worksheets, Coloring Pages, Activities, and MORE

A book by Flock University

Illustrated and written by Stacy Tate

To Julia,
Without you none of this could ever happen.

Chapter 1:
History of chickens
and how they got into our backyards

Chickens and humans have lived together for at least 8,000 years! Scientists believe the first domesticated chickens were descended from the wild red junglefowl of southeast Asia. These wild chickens have the Latin name of *Gallus gallus* .

Southeast Asia today has many countries in it. There is Thailand, Cambodia, Vietnam, Laos and more. Most people credit what is now Thailand as being the first place chickens were being used by humans.

People credit Thailand with the origin of chickens because of the red junglefowl, but there is evidence that other countries, like China, were begining to keep chickens at roughly the same time. Either way, chickens originated in Asia because it is where red junglefowl are found naturally. Countries and their borders change all the time. They always have and they always will. When we talk about history like this, we use modern names for places so you can know where to look on a map or globe and you can have an idea of what that land looked like based on climate and region. We will get more into this in a few pages. Thailand and their neighbors have a jungle climate, hot summers and rainforests.

8,000 years ago (or about 6000 BCE), humans began farming to feed themselves. Southeast Asia is known for rice and millet farms, two things chickens love to eat. The red junglefowl would get a steady supply of food; and in turn, the humans could add eggs and meat to their diet.

This is a map of Southeast Asia with Thailand in orange. Do you see China? Pictured here is just a tiny bit of China. China is actually a very big country, but the southern most part is warm and similar to Thailand, Vietnam, Laos and others in climate and vegetation.

Humans haven't always lived like we do now. We didn't always live in houses like we do or cities like we have or countries with leaders and borders like we do now. We started out as something called "hunters and gatherers" which is just what it sounds like. We traveled and lived in smaller family groups, finding food by hunting and gathering. Then we learned how to farm and grow foods like rice and wheat and keep animals like cattle and chickens. That is where the chicken's story begins, in the ancient farms of our ancestors. The chicken is something we all have in common. They spread around the world with people traveling and trading.

After awhile humans started to understand something called "animal husbandry". This is the act of keeping animals for farming, which includes breeding. Breeding is pairing up animals with certain traits to have those traits show up in their babies. For example, a hen that lays better than the others would be chosen to be the mom of the next generation of chicks. Those traits get carried on and your flock would become better and better!

Humans picked chickens for things like friendliness, feather patterns, how many eggs they laid, how good the roosters were at being protective, and many other things. After thousands of years, across many continents and cultures, we have so many different types of chickens. There isn't a book out there that lists them all!

The map on the previous page is what Southeast Asia looks like now. Countries were not always there, and they were not always called what they are today or shaped the same. Look at the map on the previous page. Now look at the map below of Southeast Asia in the late 1800's. A lot has changed. Before many places in this region were their own countries, they were ruled by other countries. It is called a "colony" to that bigger country. Cochinchina was a French colony in the late 1800's to early 1900's. Now it is part of Vietnam and Cambodia.

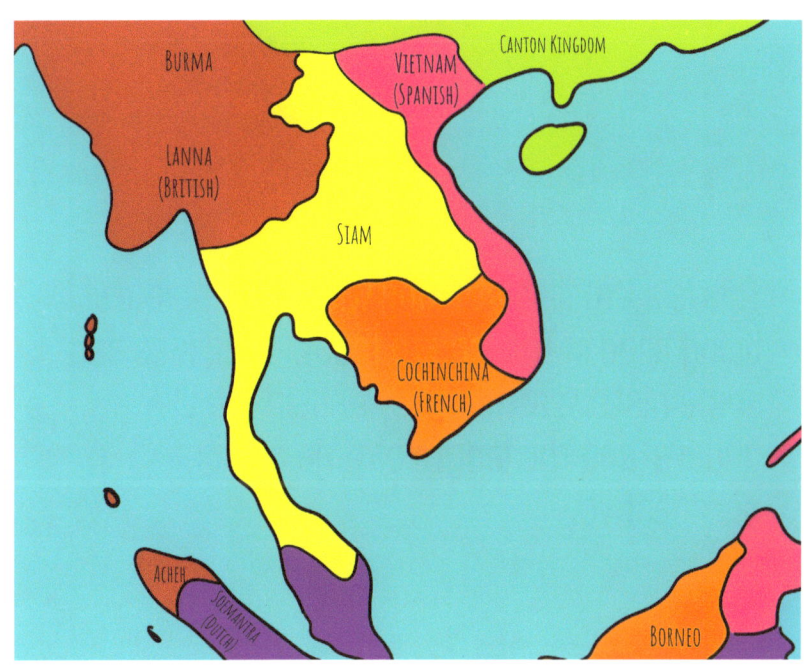

Cochin

Lays small tinted eggs
Sits on eggs ("broody")
Friendly
Big round bodies
Best as pets or exhibition birds

The Cochin is listed with the American Livestock Breed Conservancy as "watch" status.
This breed is recognized by the American Poultry Association under "Asiatic" and has a
bantam (small) counterpart also recognized.

As you can see, humans have a very long history with chickens!

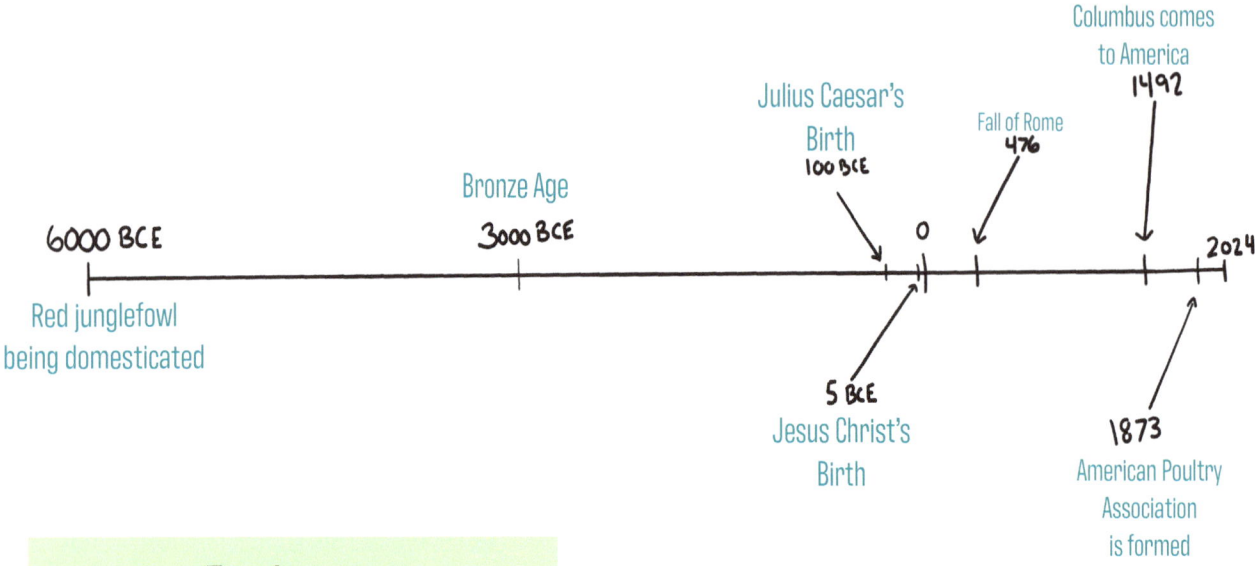

6000 BCE
Red junglefowl
being domesticated

Bronze Age
3000 BCE

Julius Caesar's
Birth
100 BCE

0

Fall of Rome
476

Columbus comes
to America
1492

2024

5 BCE
Jesus Christ's
Birth

1873
American Poultry
Association
is formed

Timelines are visualization tools to help you see the span of history.

This is a timeline starting with red junglefowl being used in Thailand to today. Timelines add some historical markers to help the viewer understand the length of time and get perspective.

Use the space below to draw a timeline of your own. Start with your parent's birth and go to today. Don't forget to add your birth and your brothers and sisters if you have them. Add whatever you want that is important to you, but remember it should be in chronological order.

These junglefowl were a far cry from the chickens we know today. As time went by and whole civilizations were being born, the chicken made it's way across southern Asia and into the Mediterranean and then into the rest of Europe. Each civilization bred it's own spin on the chicken to serve it's own needs.

It is not documented, but you can bet chickens traveled on routes like The Silk Road where travelers and traders went back and forth from home to far away lands to buy and sell goods. Even if chickens were not traded, chickens made a great travel food source. It is only recently that we have restaurants and grocery stores. Before people had to farm, cook, hunt, and gather their food. Imagine taking off from your house to go somewhere far away and not having food places along the way. Even hunting food, the animals in different parts of the world are different. You may not be aware of what to expect. Taking something small like a chicken would make total sense.

The Silk Road

The Silk Road (sometimes called the silk routes) was a network of pathways used for about 1,500 years linking the Far East to Europe. It was about 4,000 miles long! Explorers and traders would make the trek and do business along the paths. They started to call it "The Silk Road" in the 1800's because the silk that Europe used came from China on these trade routes. Silk is made from a specific Chinese moth that makes silk.

The Silk Road was first known to be used around 114 BCE. It was used until the mid 1400's. On our timeline from before, that means it was used from close to Julius Caesar's birth up unto when Columbus sailed to America! People traded silk, spices, weapons, and even ideas and stories. It is safe to say, the world would be very different without it. Do you think they probably traded chickens too?

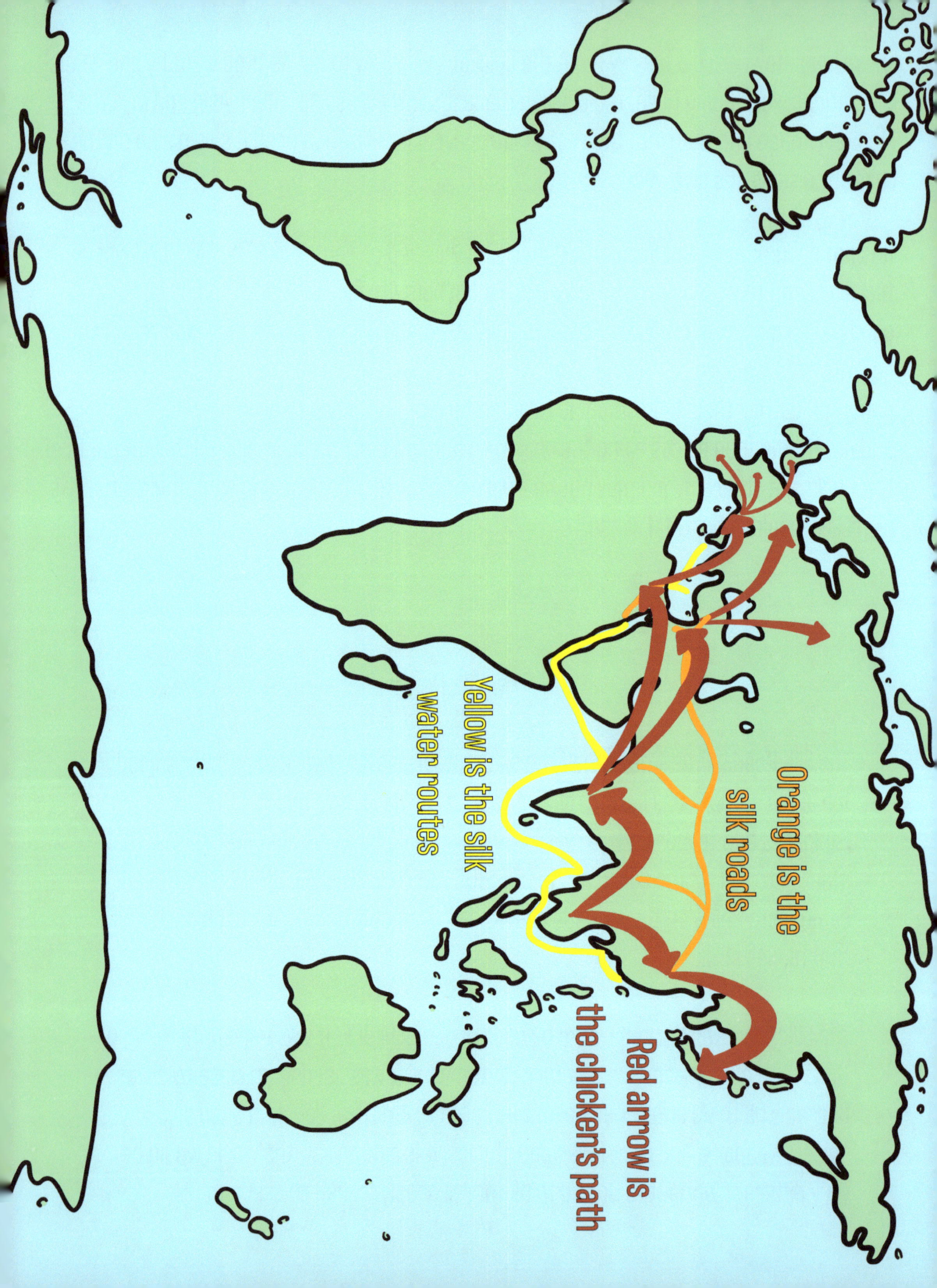

Orange is the
silk roads

Yellow is the silk
water routes

Red arrow is
the chicken's path

Ancient Egypt's Egg Ovens

Egyptians were one of the first to learn how to incubate eggs. To hatch chicks naturally, hens have to stop laying and start sitting on eggs for 3 weeks to hatch chicks. As far as we know, Egyptians were the first to figure out how to hatch the eggs artificially in large ovens like the one we can see into below.

There are some historical references from as far back as 300 BCE where Aristotle wrote about how bad the oven smelled. They didn't have open fires in the ovens, they smoldered dung. "Dung" is a nice word for animal poop. I bet it did stink!

In 1323, an Irish Friar visited Cairo and was amazed by the number of chickens they had roaming the streets. These chickens were hatched in these egg ovens. The family that worked in these hatcheries would live in the same structure and hand turn the eggs for the whole 3 weeks. The chicks were allowed to run in the center hallway, which was probably very warm, as well. The friar, named Simon Fitzsimmons, spread word in Europe that Egyptians were hatching chicks without hens and roosters. Everyone misunderstood it to be magic that the eggs hatched. Today we know it's just science.

A French traveler had gone to Cairo, Egypt and after seeing the hatching ovens, he expressed that Egypt should be more proud of the hatching ovens than the pyramids! He went into more detail about how they hatched the chickens. He described them checking the temperature by putting the eggs to their eyelids. He went back to France and tried to copy the ovens and failed. France is much colder than Egypt, and Egyptians had thousands of years of practice perfecting the art of hatching.

To imagine what these chickens looked like roaming the streets of Cairo, we only have to look at the Egyptian Fayoumi. This is an ancient breed thought to come from just outside Cairo. The Fayoumi is a smaller chicken, does not like confinement, and is most known for being disease resistant.

Egyptian Fayoumi:

Lay white eggs
Great foragers
Very good at tolerating hot weather
Stand out for being disease resistant
Is a landrace

American Livestock Breed Conservancy is studying numbers to help save this breed from extinction.
This breed is not recognized by the American Poultry Association.

The Old Stone Gate in Khan El Khalili Bazaar in Cairo has been a market since the 1300's and still a active market today.

Ancient Egypt as a whole

The hatcheries of Egypt were and still are extremely impressive. Egypt was ahead of it's time in so many ways. Most people think of The Pyramids of Giza when they hear someone say "Egypt". There also was The Great Sphinx of Giza which is a giant statue of a cat with a human head. That human is thought to be the Pharaoh Khafre who was in charge when it was built. "Pharaoh" is the word Egypt had for their leaders. The most famous pharaoh is King Tutankhamun or King Tut, as people have come to start calling him. He was the youngest pharaoh. He was 9 years old when he came into power. How old are you now? Can you imagine running a whole empire at 9 years old? Would you want to do that?

The Great Pyramid of Giza
The largest pyramid in Giza is called The Great Pyramid. It has 4 sides that are each 755ft long (230 meters) and is 455ft (138 meters) tall.

Hieroglyphs

Hieroglyphs were the way ancient Egyptians wrote down stories and history they thought important. They used pictures and shapes to express their words and thoughts, instead of letters like we do in Western languages.

We could not read hieroglyphic writing until the Rosetta Stone was found in 1799. The Rosetta Stone had two other languages on it other than hieroglyphics, one being Greek. Since we knew Greek, we could translate the hieroglyphs. That is how the Rosetta Stone was the key to reading hieroglyhs.

On the next page, take note of where the city of Rosetta is located.

Partial map of Ancient Egypt

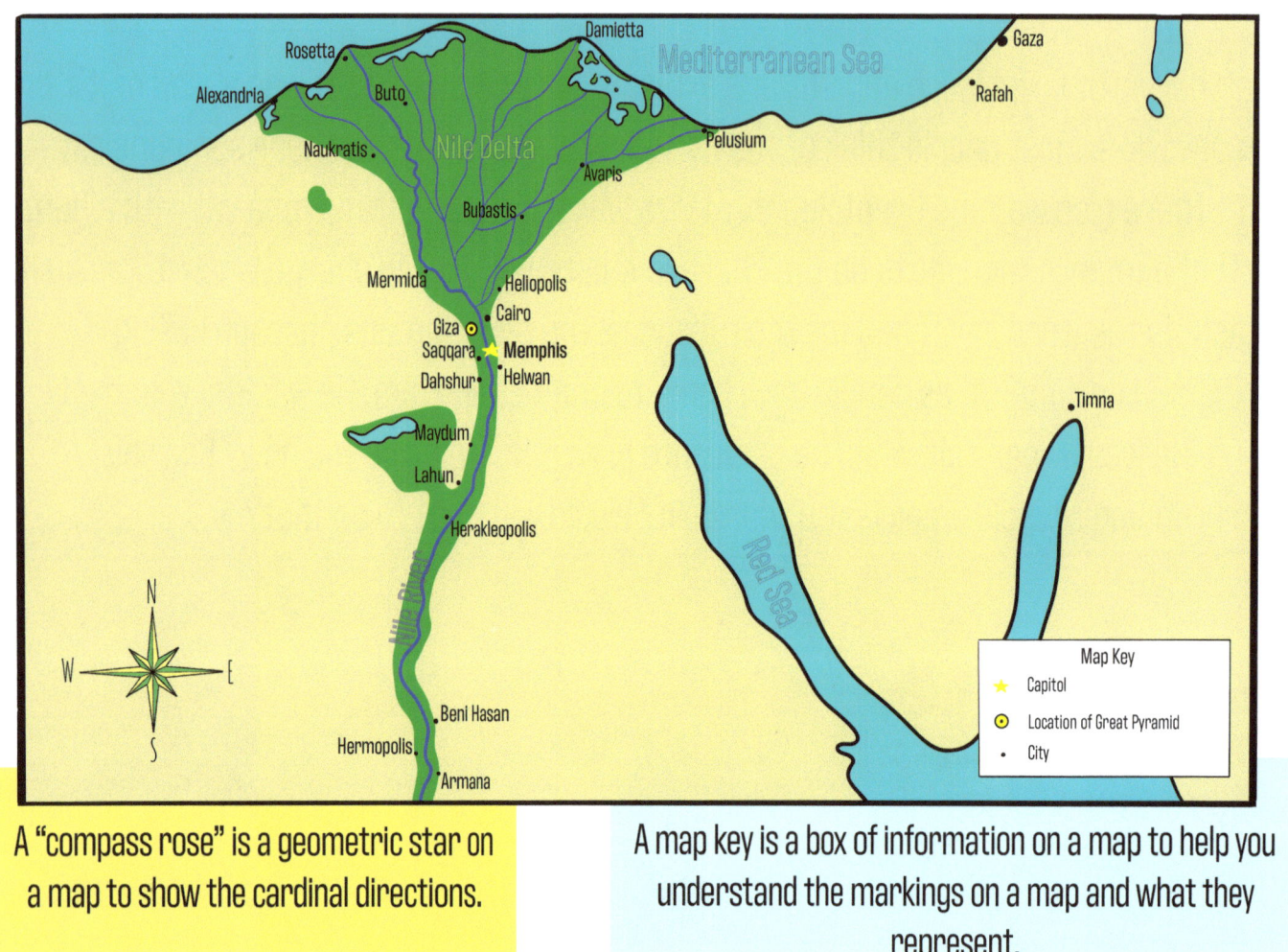

A "compass rose" is a geometric star on a map to show the cardinal directions.

A map key is a box of information on a map to help you understand the markings on a map and what they represent.

Fact about the Nile: it flows north!
There are about 250 rivers in the world that flow north.
The Nile is the most famous.

This map shows part of Ancient Egypt. See the green area that looks a bit like a fan or flower? That area is called the Nile Delta. A delta is a type of landform. It is fan or triangle shaped and happens when a river sediment (that is a big word for sand and mud) gets pushed from a flowing river into a larger body of water that does not carry the sediment away. The Nile Delta is named because it is a delta that was created by the Nile River. We have a famous delta in the USA. Do you know where it is? Louisiana! The Mississippi River has a delta that comes out below New Orleans.

Most of Egypt is desert and sandy. Along the Nile River is green and lush because the water from the river waters vegetation. Notice how many cities are along the river. Do you think Egypt would have been such a big and advanced civilization without the Nile?

Check out some YouTube videos on ancient Egypt:

"Ancient Egypt For Kids" by edyout00
"All About Ancient Egypt for Kids!" by Twinkl Teaching Resources
"Who Were The Ancient Egyptians?" by 60 Second Histories

ROMANS AND ANCIENT SACRED CHICKENS

Chickens played a huge role in human history and not always on our plates. In the Roman Empire, they served in a very different way. Romans had a belief that they called Pullarius, or as we call them today, "The Sacred Chickens". These chickens were kept to predict the future. All extremely important decisions were made... by chickens.

Sacred chickens were housed in a special kind of coop called a Pullarium. A priest would care for them and observe their behaviors, translating their actions into predictions. Important decisions were made this way, so in a way, it was chickens that built Rome.

Before battles, the sacred chickens were not fed. Just before deciding when to attack, these chickens were given food. If they ate it, then that was a "yes" and approval from the heavens to go ahead with the plan. If the chickens did not eat, then that was a "no" and a sign that the plan would fail.

There is a record of a Roman commander ignoring the sacred chickens' prediction of the battle he wanted to have in the First Punic War. He lost that battle so badly that he lost almost all his ships and men. When he arrived back in Rome, the people were so upset with him ignoring and disrespecting the chickens that they had him put to trial and forced to pay a huge fine.

Chickens were also a food source for Romans, as well. Romans loved food and created new ways of cooking chicken. They created new recipes and bred their chickens to be big birds with lots of meat. Some of these recipes are still around and loved today. The breed we have today that is credited with being developed by the Romans is the Dorking. It is a big chicken that is best known for it's meat flavor and calm temperament.

Marcus Terentius Varro was one of many ancient Roman writers who wrote about chickens and how to properly care for them. He stands out as the writer who wrote the most about chickien keeping. Varro was born in 116 BCE and died in 27 BCE. He wrote about how to make sure hens hatch healthy chicks, how many chickens to keep in a flock, how to build coops for best results, and more. So much of what we still hold true today, he wrote over 2000 years ago!

Romans loved chickens so much they decorated their homes with mosaics of chickens, their businesses, and even their coins had chickens on them.

To the left is a Roman fast food stand that they have uncovered in Pompeii. You can search the internet for more pictures of the real one.

Aquaducts carried water from one place to another. Think of them as manmade rivers. The arches are called "Roman arches" and have proven to be extremely strong and many last still to this day. Architects and engineers use arches all the time for strength when building.

Dorking

Medium to large white eggs
Used for eggs and meat
Gentle and not aggressive
Stands out for meat flavor
Has a bantam (small) size

Livestock Breed Conservency listed as "Watch" status.

American Poultry Associated has this breed listed as
English class.

Roman Empire As A Whole

The Roman Empire is one of the most famous ancient empires. It was not the largest or the last, but it was one of the most influencial. On the page before this one is a picture of an aquaduct and, though many have been lost to time, there are plenty that still remain. Some still are being used! There are Roman roads that still exist and are being used. Other things we use from Romans are Roman numerals, our legal system and our political system.

The Roman Empire in red, at it's largest in 117 CE.

Rome was founded in 753 BCE by two brothers, Romulus and Remus. Legend has it that both brothers were placed in a river as babies by their uncle, the king of nearby Alba Longa. A farmer found and raised them. When then boys grew up, they killed that king and then Romulus killed Remus, taking control of the city and naming it Rome after himself.

Rome spread from being a city state to being a full blow empire by expanding it's territory by military action. As you can see in the map above, at it's largest size, the Romans had most of Europe, the Middle East, and the northern African coastline.

The Roman Empire officially ended in 476 CE. That is roughly 1000 years the Roman Empire was active and influencing human history. The Roman Empire fell for a number of reasons, but one of the biggest reason was government corruption. In the end, German Goth Odoacer defeated the Romans and marked the end of The Roman Empire.

Julius Caesar is probably the most famous Roman leader. He was born July 12, 100 BCE to a rich Italian family. He was in the Roman army and advanced to be governor of Spain, which was Roman at that time. He went back to Italy fighting a man named Pompey for rulership. Caesar won and declared himself dictator for life. He was assassinated by members of the senate in 44 BCE who did not want to be dictated by a life long ruler.

As Romans were spreading their empire, they built roads and aquaducts in the cities they conquered and made the cities even better for more people to live there. They created school systems, legal systems, and government systems that we have built our current societies upon. They created ways to deal with sewage and created baths, keeping sickness down. These things were advanced for their time. After Rome's collapse, Europe went into the dark ages, forgetting and not using the advancements the Romans gave them.

When cultures are thriving, people have more time to relax or entertain themselves. They are not struggling every moment of every day to just survive. These cultures need entertainment and art. Romans were famous for colloseums where they would put on shows of past battles and have slaves and gladiators fight each other and animals for entertainment. With an empire spreading to what is now 3 continents, they had access to many types of animals and people. They had passages and cages through the giant colloseums with trap doors to have dramatic preformances. They even could fill the floors in with water and re-enact naval battles. The most famous colloseum is the one in Rome and it is still there today and open for people to tour.

There were many cities, as you can imagine in an empire so big. One of the most famous in Italy, outiside of Rome itself, was called Pompeii. This city was placed at the base of a volcano named "Vesuvious", and it's suspected they didn't realize there was a danger. In 79 CE, Vesuvious errupted and left the city under 20 feet of ash. There were earthquakes before the erruption, but at the time, they didn't realize that it meant the volcano would errupt. Pompeii was an important farming and trade city for the empire. They had fertile farm land from the older erruptions and it was a port city so they could fill ships for trade. Today, historians are still discovering more from Pompeii. Most famously, everything covered in ash was preserved, including the people's poses and what they were doing. This gives historians a big insight into daily Roman life. It also is a place you can visit, even now.

 LEARN MORE ABOUT ANCIENT ROME BY WATCHING THESE YOUTUBE VIDEOS:

"ANCIENT ROME 101" BY NATIONAL GEOGRAPHIC
"ANCIENT ROME FACTS FOR KIDS" BY FUN FACTS KIDS TV
"FUN FACTS ABOUT ANCIENT ROME DAILY LIFE" BY THE ANCIENT LIBRARY

HERE IS A VIDEO OF A RESTUARANT IN POMPEII. IT IS PRETTY COOL TO SEE HOW 2100 YEARS LATER AND OUR FAST FOOD PLACES LOOK VERY SIMILIAR.

"IL TERMOPOLIO DELLA REGIO V" BY POMPEII

Regional Breeds and Landraces

As chickens spread around the world, they looked less and less like the red junglefowl that started it all. Chickens were starting to be found in colder climates about 1000 years ago. They were starting to change and adapt to their new environments.

Chickens were being kept in Russia in the 9th century (which is 800 CE). Russia is a lot colder than Southeast Asia. Those chickens probably didn't look much like red junglefowl anymore. They probably looked a bit more like Russian Orloffs. That is a breed of chicken that has lots of feathers, including more on the face to make it look a little bit like a beard. This helps them stay warm in cold weather. That is something needed in Northern Europe! It is a perfect example of how chickens changed to meet the needs of the environment and cultures they were kept in.

This Russian Orloff rooster looks ready for winter.

Russian Orloffs were named after a Russian Count who helped develop the breed and died in 1808. Orloffs lay small to medium tinted eggs, are cold hardy and great foragers.

They are listed with The Livestock Conservancy as "Threatened" and they are not recognized by the American Poultry Association.

Chickens were kept in regions and villages that were sometimes isolated from trade. Often those places ended up with what is called "landrace" breeds. This is a term used for a breed of chicken that came about through natural selection and not human selection. A good example of a landrace chicken is the Sulmtaler. The Sulmtaler is a breed from the Sulm Valley in Austria. They do well in cold weather and lay tinted smaller eggs. To understand how isolated these Sulmtalers were, look at the maps below.

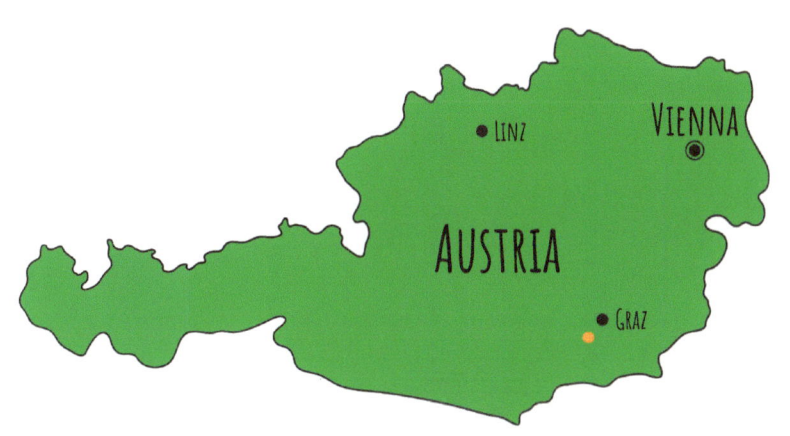

Sulm valley is the orange dot. Until recently, it was the only place you could find Sulmtalers.

Look for Austria on the map.
Do you see how small an area Sulmtalers were found in?

CRITICAL THINKING

In today's world, we can go online and buy landrace birds from all over the world. What do you think this means for those breeds? Will it help save them to have them in places they never lived before? If an Austrian Sulmtaler lives in Oklahoma, is it still a Sulmtaler? What do you think? What if people let their landrace birds breed with other breeds in their yard and grow their chicks? What is your opinion? Does this mean the end of landraces with the whole world able to buy, sell, and trade with each other? Remember, the definition of a landrace is a breed that came about from natural selection in a specific environment, not human interference and selection in a wide range of environments.

_____

_____

_____

_____

_____

_____

_____

_____

_____

THE OLD AND NEW WORLD

Now we know how chickens got to Europe from Asia. European countries like England, France, and Spain traveled the world with ships and spread chickens even further. This time period was called The Colonial Period and these countries were sailing the world, expanding their territories and taking chickens with them. If we go back to the cochin, remember it came from Cochinchina? That was a colony governed by France. Here in North America, we were colonized by Spanish, French, and English. The Spanish occupied Florida and Mexico, France occupied Canada and Louisiana, England occupied what we call "the 13 colonies". They brought their "old world" chickens to The New World, a name given to America, and it didn't take long before we started to breed the chickens to be what we needed them to be on North American farms.

Chickens came to America with the Europeans on their ships.

Jamestown, Virginia is credited with being the first English settlement. St. Augustine, Florida was settled even earlier by the Spanish and has the USA's oldest schoolhouse. You can visit it still today!

The American breeds of chickens are some of the newest breeds overall, but they come from old stock crossed and bred to be what the farmer needed or wanted out of birds. Many American breeds are "dual purpose" which is a term used for a breed that is good in meat and in eggs. Imagine living in early colonial America, what would you want your chickens to be like?

Wyandottes and their colonial coop.
Wyandottes are named after a Native American tribe of the same name.

Some of the first American breeds were Plymouth Rocks, dominiques, and Wyandottes. The birds needed to be hardy and in the northern parts, they needed to be cold hardy. Winters were always a challenge for most of the colonies. In Florida, we are unsure what breeds they had, but they almost certainly were game birds. Still today there are wild chickens living freely in the deep south, ancestors of colonial chickens. Key West, Florida is known for their wild chickens.

By the mid 1800's people were breeding what they had and calling the breeds whatever they wanted. In a lot of ways, it was chaos. Trading and traveling was easier than ever and the genetics of chickens was starting to get all mixed up. In Europe and in America chickens were becoming something people were falling in love with. They called this "hen fever", and it changed how we breed and keep chickens. Hen fever had people buying and collecting breeds they thought they wanted.

In England, the popularity of chicken shows had been rising since the outlawing of cockfighting in 1849. The first official show was in 1845 in London. People loved showing and seeing chickens of all kinds. Even Queen Victoria attended and entered her cochinchinas chickens. Charles Darwin loved to be a spectator. Chicken shows were a big deal!

The very first American competition was supposed to be in Boston, Massachusetts in 1849, but there was no set standard and no way to judge the birds. They could not compete due to not having a way to judge the birds, so people just showed off their chickens. They had 1,400 entries! That is a lot of people wanting to show off their chickens!

P.T. BARNUM HAD A NATIONAL POULTRY SHOW IN 1854

The problem became clear quickly. There was no set guidelines for what was a cochin or a dominique or a crossed up barnyard bird. The hen fever sweeping over Britian and America was further fueled by Queen Victoria and her flock of cochins. Writings from that time describe a bird with feathered feet tall enough to eat off the top of a barrel, though it did not specify what size barrel. Between everyone wanting fancy chickens and the lack of order to all the varieties, organizations started to be formed to define what each breed was supposed to be.

In the US, the American Poultry Association was formed and with it a book called "The Standard of Perfection" was published. They still publish this book today. It is sometimes referred to as "the chicken breeder's bible". It lists the breeds the APA recognizes and describes every detail that bird should have, both standard size and bantam (or smaller version). This allows for definition and uniformity, and ideally preservation of those breeds.

Dominique

Lays medium tinted eggs

Good forager

Cold and heat hardy

Friendly

Noted as America's first chicken breed

Listed on the "watch" list with the Livestock Breed
Conservancy

Recgognized by the APA in the American catagory. Does
have bantam verson, as well.

Paul Revere House in Boston, Massachusetts was built in 1680 and is the oldest structure in downtown Boston today. Paul Revere was a silversmith during the American Revolution and is famous for his midnight ride where he warned other American colonists that the British were coming.

In 1948, there was another type of contest. This one was called "Chicken Of The Future" and it was a competition to see who would make the fastest growing white meat chickens for market. It was put on by the first huge grocery market and the world was forever changed. The store was called A&P and put a lot of small grocery stores out of business. They put on this competition to have farmers breed them a chicken that they coud sell faster and make more money. The winner was Arbor Acres of Connecticut and we still eat descendants of those birds today.

With the rise of the commercial food industry, we have had a loss of genetic diversity. The chickens that fed the world up until around 1950 were from people's towns and backyards. There are some people who try to keep older breeds going. Breeds recognized by the APA before the mid 1900's are called "heritage" breeds. These breeds grow slower and breed true, naturally. There is an organization called The Livestock Conservancy that is dedicated to helping save these heritage birds. There are breeders and farmers all over the US doing their part to save these breeds from extinction, as well.

Find full up-to-date information on hertiage breeds (and breeders)
on livestockconservancy.org

Wyandotte

- Named after a Native American tribe
- Dual Purpose American breed
- Lays large brown eggs
- Large round bodies
- Graduated off of endangered list in 2016
- Originally called "American Sebrights"
- Cold hardy
- First breed to be recognized by the APA in 1883

Yokohama

- Ornamental breed named after the port they exported from
- Lays a small amount of small tinted eggs
- roosters tails grow 3 feet a year
- British and German breeders refined the breed from the original Minohiki breed of Japan
- Listed as "threatened" by livestockconservancy.org
- added to the APA in 1981

Buckeye

- American dual purpose breed
- developed by a woman in the late 1800's and was accepted by the APA in 1905
- Great at foraging, does not do well in confinement
- Lays large brown eggs
- cold hardy
- Known to hunt mice, roosters known to make unique sounds in addition to crowing
- On the "watch" list with livestockconservancy.oeg

Sulmtaler

- A landrace from the Sulm Valley in Austria
- Recent years has been imported to the United States
- Dual purpose breed that is recognized in European poultry clubs, but not by the APA here in the US
- Very cold hardy, does not do well in heat
- Lays large cream colored eggs
- Not studied by the livestockconservancy.org

Appenzeller Spitzhauben

- Swiss breed believed to originate in 1500's
- Lightweight and flies well, does not tolerate confinement
- Unique comb
- cold hardy
- Lays medium to large white eggs
- Not recognized by the APA, it is recognized in Europe
- Listed under "watch" by livestockconservancy.org

Sussex

From Sussex, England where they historically have been known for meat

Lay light to medium brown eggs

Very calm

This breed has a color called "Coronation" which is white with lavender neck and tail and is very rare.

Listed as "Recovering" with livestockconservancy.org and recognized with the APA in English class

Japanese

Asian ornamental breed that is a true bantam, meaning there is no large version.

Lay tinted or cream colored eggs

Some are born with long legs and those should not be bred, especially for competition

Not good for cold climates

They are good foragers but all must be protected from predators more than larger chickens

They are listed as "threatened" by the livestockconservancy.org and are recognized by the APA

Barnevelder

Dual purpose chicken from Holland

Lay large dark brown eggs

Come in "double laced" coloring pattern

They have a personality that is calm and friendly

They are cold hardy, not so good in extreme heat

Not studied by livestockconservancy.org and is recognized by the APA

Delaware

American dual purpose breed

Lays jumbo brown eggs

Originally created as a broiler, or meat breed

Very hardy in heat and cold

Good forager

Calm and friendly

Listed as "recovering" with the livestockconservancy.org and is recognized by the APA

Sicilian Buttercup

Very unique buttercup comb

Lays white eggs

Heat tolerant, they do not do well in cold climates

They have a flighty and chatty personality

They are listed as "threatened" by the livestockconservancy.org and they are also recognized by the APA in the Mediterranean class

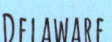

Science explains how things continue on through life cycles. All humans start out as babies and grow into adults and then have babies of their own. Chickens are not much different. In this chapter, we will learn about the chicken's life cycle and explore the different stages of life a chicken goes through. Below is an example of a chicken's life cycle to help you get familiar.

Chicken's average lifespan is 3-10 years. The oldest chicken recorded was "Muffy", a Red Quilled American Game hen who was 23 years and 152 days old!

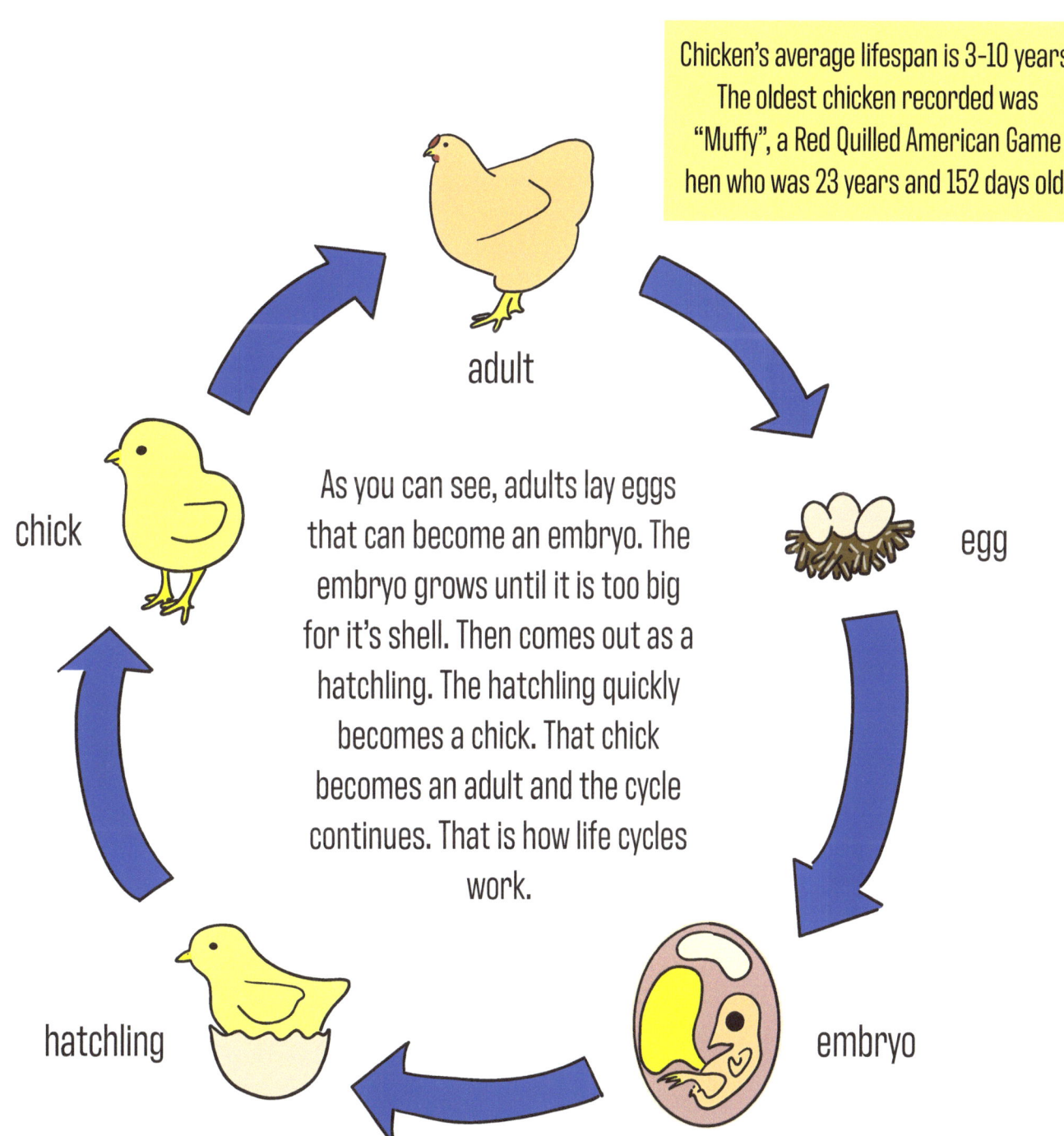

adult

egg

chick

As you can see, adults lay eggs that can become an embryo. The embryo grows until it is too big for it's shell. Then comes out as a hatchling. The hatchling quickly becomes a chick. That chick becomes an adult and the cycle continues. That is how life cycles work.

embryo

hatchling

EGGS AND EMBRYOS:

The adult females, called hens, lay eggs and naturally can incubate the eggs. "Incubate" means maintaining a temperature that allows the embryos to develop and grow. Eggs need two things to start to develop embryos. They need to be fertilized and need to be warmed up to 99.5F. If you have a rooster with your hens, chances are good they are fertilized. Once the hen has a good amount of eggs in her nest, she may become "broody". This word literally means they are moody or preoccupied, which is exactly what happens. Broody hens can be quite protective of their nest and eggs. Broody hens collect eggs, and not always their own. It is not unusual for a broody hen to collect the eggs of other hens. She can take the eggs under her wings and carry them back to her own nest. They get quite greedy about eggs during this time! She will move the bedding around her in a way to keep warmth in and will sit on the eggs for 3 weeks. Occassionally, she will move them around and check on how they are doing. She will only get up to eat and drink once a day, making sure the eggs are covered and safe before she gets up to take care of her own needs.

100   ← 99.5 F

BROODY HENS:
Always respect broody hens. They are willing to do anything to protect their eggs. Even the sweetest birds will fight anything to protect their nest so they can hatch their babies.

It takes an egg 21 days for an egg to develop into a hatchling. The conditions need to be perfect for this to happen. Like we talked about already, the temperature has to be 99.5F, the humidity (that is the amount of water in the air) has to be right, and they also need to be turned. One of these things going wrong can ruin the hatch.

You should check eggs on day 6 or 7 by candling and take out eggs that aren't developing embryos. Broody hens do this too. Science has not figured out how she knows, but she knows which eggs are developing and which are not. She will roll eggs that won't hatch out of her nest to make sure the other eggs stay clean and healthy.

CANDLING:
Candling is a way to check on egg development as you are incubating. You put a flashlight or candler to the egg and you can see if there is a embryo inside! Remember to be gentle, you don't want to break the shell!

HOW EGGS LOOK WHEN YOU CANDLE THEM:

not fertile

early death

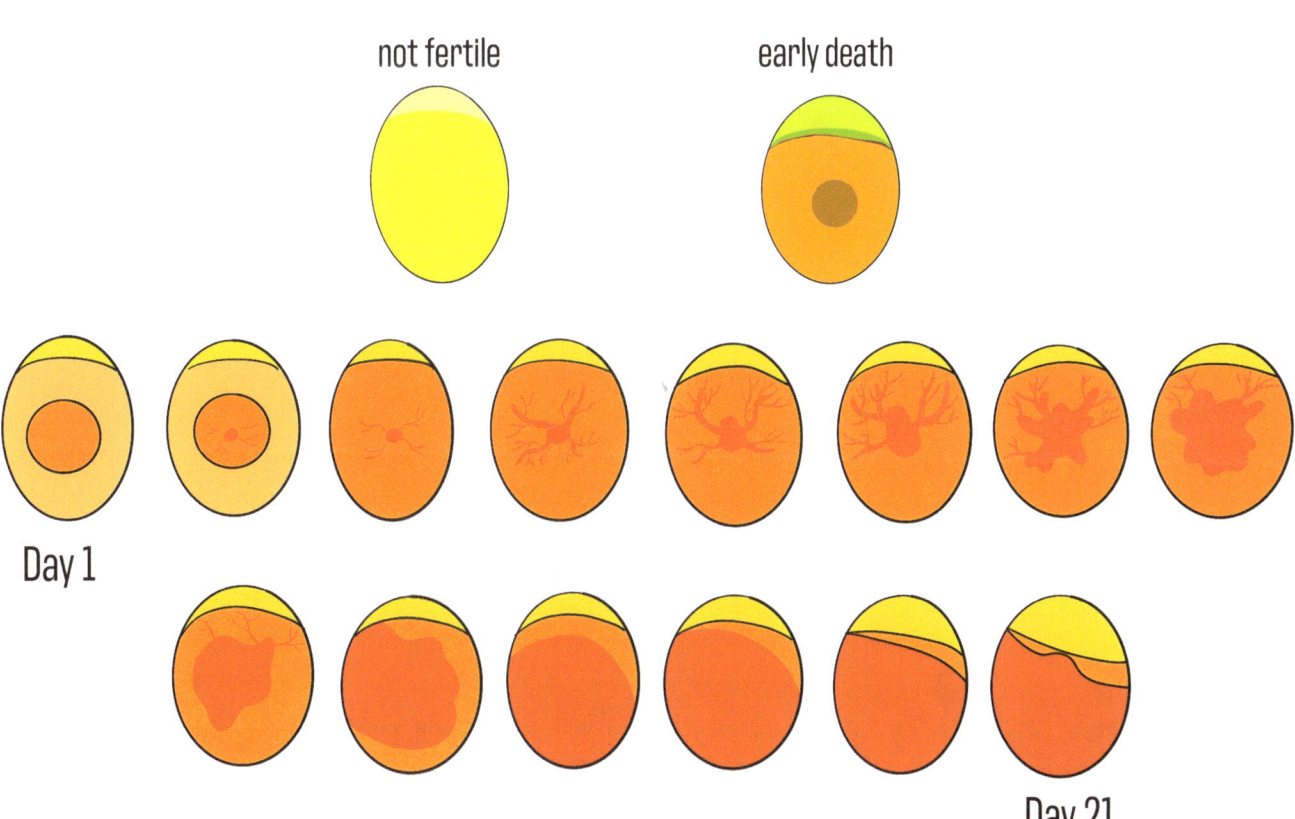

Day 1

Day 21

As the egg develops, we get the embryo stage. This is a stage of life we all go through. You too! It is hard to explain simply, but the best way to describe it is that there is life inside the egg, but it isn't exactly a chick yet. It is growing in the shell for 21 days and by the last day it needs air and is a chick. It knows what to do. Each baby has a little point on their beak called an "egg tooth" and it is there just to help the baby break the shell so they get an airhole. This is call a "pip". As the baby starts to move it's head around, it will then keep breaking the shell and this is called a "zip".

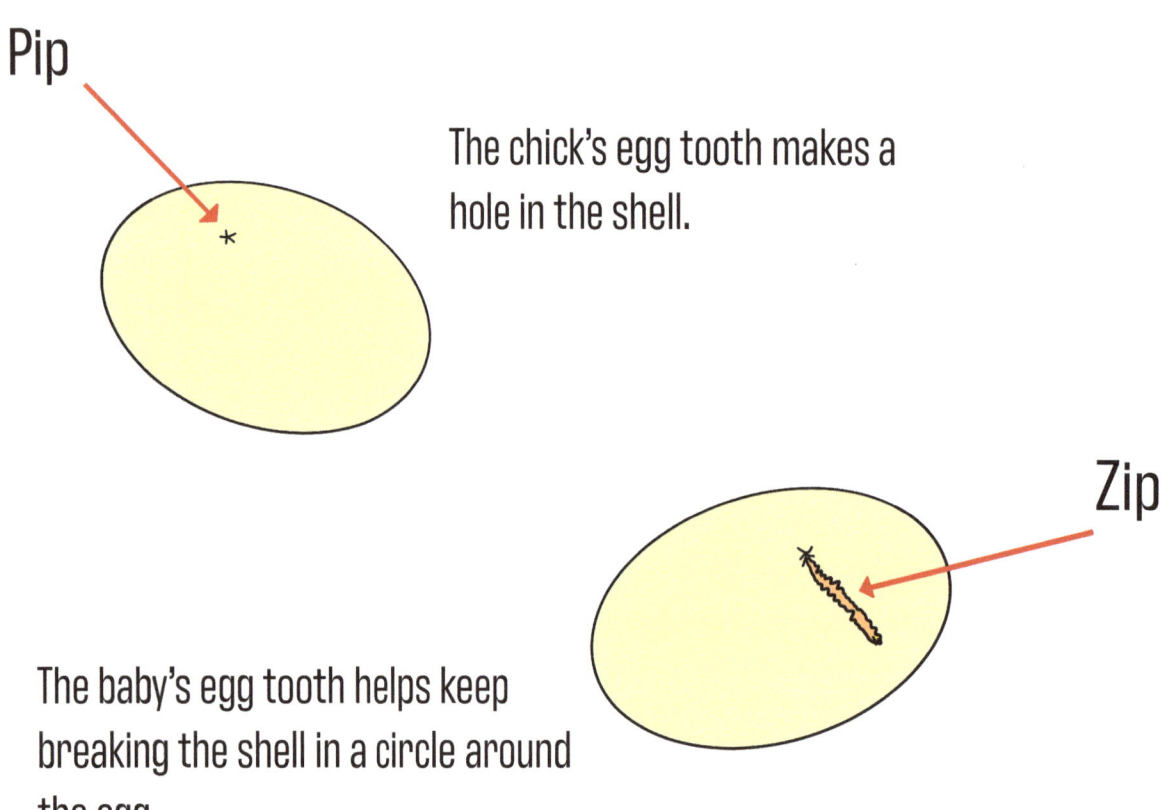

Pip

The chick's egg tooth makes a hole in the shell.

Zip

The baby's egg tooth helps keep breaking the shell in a circle around the egg.

When the egg gets fully zipped, the baby can pop out. The hatchlings are wet and need to stay in the hatcher until they fluff out. They use this time to gain strength in their legs too. It is normal for them to be a little wobbly in the first few hours.

CHICKS:

Chicks are babies that have dried after hatching, usually in the first day. If you are raising them and not the mama hen, this is when you can put them in the brooder. A brooder is a safe, warm place to raise chicks. Different people make brooders in different ways, but they all need to be free of drafts and have enough space for the chicks to grow and move in and out of heat. They need 95F the first week. Then each week after, they need less heat until they don't need it at all. Some people use heat lamps and some people use heat plates. These are bought at feed and farming stores.

Chicks need a food called "Chick Starter" which has all the stuff they need to grow big and strong. Think of it as baby food. After a few weeks some people then start to feed their chicks "Chick Raiser" which is very similar but meant for slightly older chicks. It is ok to keep them on "starter" if you can't find "raiser", though. As we go through the next pages, you will start to see, different ages get different types of feed.

Here are two different types of brooders. The most important thing is being safe with the heat source and keeping the babies healthy.

PULLETS & COCKERELS:

Just as children grow to become teenagers, chickens also have a stage of growth between chick and adult. This stage might be called things like "started birds" or sometimes "feathered out young birds". The technical correct term for a teenaged female is "pullet" and a teenaged male is "cockerel". This is the age when male birds start to look different than female birds.

This age group is going to be fully feathered and able to be outside. They still need lots of protection, though. They are small and they are learning. If you have this stage being raised by a mama hen, they would be getting braver and further away from her when they are out foraging for food. Pullets and cockerels should get "chick raiser" feed, if you can find it in your stores.

People keep pullets and cockerels in tractors and coops. The main thing is to make sure they are safe from predators and healthy. Tractors are housing for chickens that is open to air and can be moved very easily due to them being so light. Often people move these daily by hand. Coops are houses for chickens that typically are not moved, but sometimes people build them with wheels, so they can be moved periodically with a farming tractor or truck.

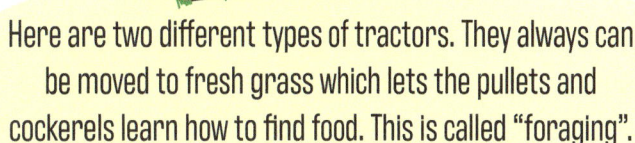

Here are two different types of tractors. They always can be moved to fresh grass which lets the pullets and cockerels learn how to find food. This is called "foraging".

ADULTS:

Chickens become adults when the hens start laying eggs. Typically, this is about 5 months of age. The boys have been crowing for a little bit too, though they start a bit early. They will be almost full sized by this point. They still should be as tall as they will be but they will still fill out in muscle a bit more. The girls are now called "hens" and the boys are now called "cocks" or "roosters".

When the hens start laying eggs, they need calcium in their diet. Calcium is a mineral that helps them make egg shells. It is the same stuff that makes bones strong. Your bones and chicken's bones need calcium. They make a feed just for laying hens that has calcium added. It is called "layer" and comes in pellets or crumbles. If the chickens are kept for meat, they get a feed called "meat bird" that is very high in protein, which is what muscles need to grow, just like you!

Higher in calcium

Higher in protein

Adult chickens are usually kept in coops when people own them for themselves. Some people raise chickens in huge chicken houses for companies that then sell the chicken or eggs to grocery stores. These are two very different ways to keep and raise chickens. People who keep chickens for themselves or small farms that sell to the public usually sell eggs and meat or sometimes just eggs. People who keep chickens for big companies typically do one or the other; eggs or meat. Those people have their industrial chicken houses set up for whichever product they are selling. For years those were all indoor only, but there has been a move to get away from that in recent years. You may go to the grocery store and see the terms "pasture raised" or "free range" indicating that the chickens get some or all of their time outside in the sun.

There are many ways to build a coop, you just need to give the chickens enough room to keep them healthy and happy, keep them safe from predators by building it strong, and make sure they have good light and ventilation. Those are the only rules. The rest can be as creative as you are. One family made a whole chicken sized Old West town. Another family made their coop look like a UFO. What would you like to make a coop look like for fun?

I WOULD LIKE A COOP TO LOOK LIKE _____

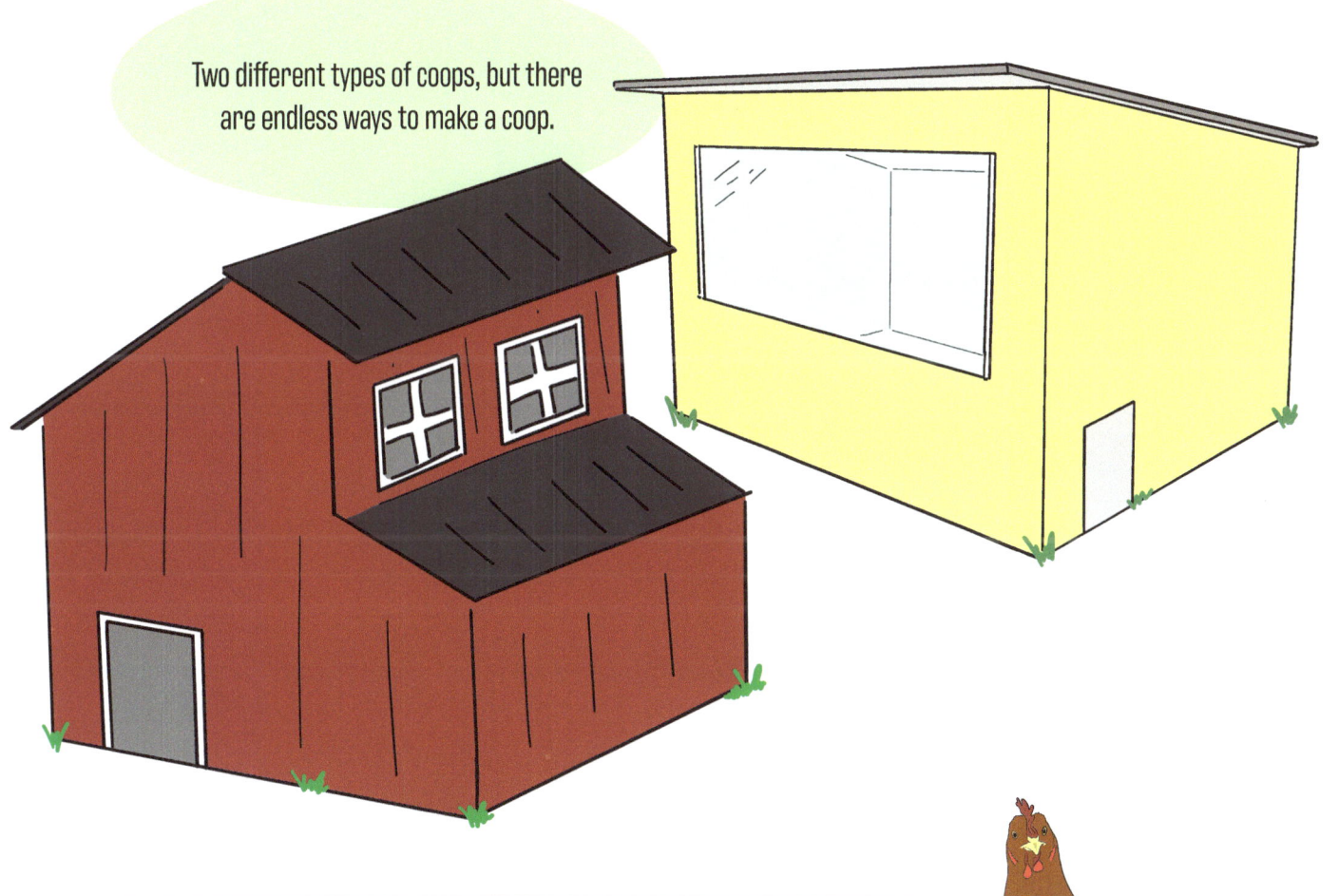

Two different types of coops, but there are endless ways to make a coop.

KNOW YOUR LABEL!

"Free Range" on grocery items means the chickens get at least half their time outside on the ground.
"Pasture raised" means they get total freedom to come and go from inside and out.
"Cage Free" means they are not kept in what is called "battery cages" where they have hens that lay in stacked cages. Lots of people find it mean to keep birds in small cages all their lives and will happily pay extra to know the hens can move around. This does not mean they get to go outside, though.
"Orgainic" on groceries means the food is free of pesticides, antibiotics and fertilizers. With eggs, it also means they are cage free.

Now we will talk about how to best take care of chickens. All living things need nutrition, which is a big word for food. We talked about how each life stage gets a different type of commercial food, but we didn't talk about how to feed them.

There are different types of feeders for different life stages too! There are also different types of waterers. When you are starting out with chickens, this can seem confusing, and quickly overwhelm you with choices. We are going to learn all about what makes the best equipment for some common situations. We are going to be learning about what size and type of coop or tractor is needed for different purposes and so much more.

To the left are some adult feeders and above are some chick feeders. Chick feeders are usually designed to only allow the head to fit in. The adult feeders are much too big for little chicks to reach in safely.

The best feeder and waterer for your situation matters. Just like we humans need good food that is not spoiled or moldy, chickens too need good food that isn't spoiled or moldy. The key to that is how you feed them and what feeder you are using. Just as you wouldn't drink your drinks out of a trash can, your chickens don't want to drink their water out of something dirty and maybe even the wrong size for them.

The question is then, "How do I know that I am getting the right feeder and waterer?" The right one will be different for every flock.

The best thing to do is to imagine you are the chicken. Can you reach the food and water? Does it seem like the food and water are clean? If you were the chicken, would this be enough food and water for you to be happy and healthy?

WHICH ONE WOULD I LIKE BEST IF I WERE A CHICKEN?

The things you need to think about with waterers:
Can the chickens reach to be able to drink?
Will they have enough until I fill it again?
Will it stay clean throughout the day?
Is it in the sun where it will get hot and grow algae?
Can they knock it over?
Is it safe or could they fall in it?

The same thing can be done for feeders. Imagine you are the chicken, how would you want your feeder to be set up? There are a few other things to think about when it comes to feeders. Feed has to stay dry. This is sometimes a challenge for some chicken people. When chicken feed gets wet, it can mold quickly. Chickens who eat mold can get sick and even die. We surely don't want that!

Another thing to think about with chickens is that they really love to play with their food. They will toss it out and spill it all over the ground if they get the chance. When picking out a feeder right for your flock, remember they will try to sling their food out with their beaks. Some people just put their feeders on a surface they can keep dry and clean. Some people use concrete pads or rubber stall mats. You just want to keep the food always clean and dry.

As you look around for a feeder or DIY feeder idea that works for you, think of the same things as the waterer.
Can the chickens reach?
Will it be enough food until it gets filled again?
Will it stay clean and dry?
Can they knock it over?

To the right is the worst set up. The feed is out in the weather, not covered at all, all over the ground and getting soaked. This is exactly what you do not want.

The same thing goes for how you store food before you feed it. You need to make sure it is always safe from getting wet and safe from animals getting in it. Lots of people like trash cans or buckets with lids.

It is not important what you keep feed in, just that it is safe from animals like mice getting in and that it stays dry. Once your feed is contaminated, it has to be thrown away so no animals get sick eating it.

There are a lot of things that can make a chicken not feel good, just like you and me. Just like us humans have to eat healthy and stay active, wash our hands and keep our houses clean - chickens need the same. Even when we do all these things, sometimes we still get sick, right? Well, again, chickens are the same. Sometimes, dispite us taking great care of them, they get sick. Now we will explore how to tell if your bird isn't feeling well so you can tell an adult to look closer.

If you have a chicken that is not well, it will try to hide it as long as possible. If they are really sick, they will look "poofed up". They tuck their head into their shoulders and their feathers are poofed to make them look more round. They might have pale skin or sunken eyes. They will look something like this:

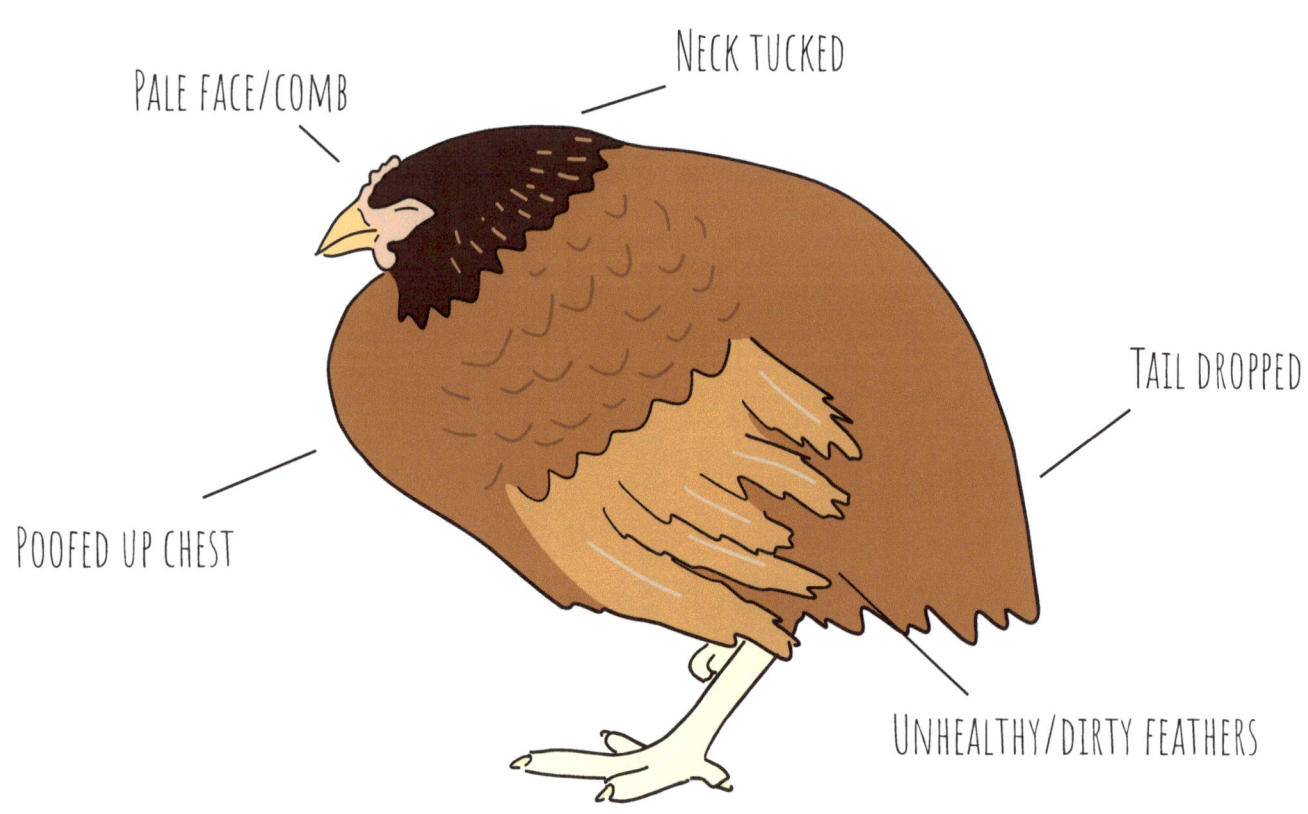

Before a chicken gets to the point like the picture on the previous page, there could be other signs. Sometimes they look fine but they are sneezing occassionally. Sometimes they have a swollen spot on their foot that is infected. Sometimes a bit of a runny nose is the only thing you can see. It is important to pay attention to your birds when you take care of their food and water. The sooner you can catch them not feeling well, the easier it will be to make them feel good again.

One really common thing to look for is a chicken scratching itself a lot. They could have mites, which are tiny bugs making them itch. Another common thing is something called "fowl pox". You get chicken pox as a human, and chickens get fowl pox. It shows up as itchy bumps that turn black and irritate them to the point where they may stop laying eggs, because they can't stop itching. Today we have the chicken pox vaccine so you may never have experienced how itchy it can be, but maybe your parents or grandparents can tell you what they remember.

Dry pox to the right in 3 stages: first is small white/yellow bumps. Then they get larger. Then last they turn black. These are now scabs. They may leave scars later.

TURN TO BLACK SCABS

GROW IN SIZE

SMALL WHITE BUMPS

It is important to understand, these are two totally different viruses and you cannot get fowl pox from your chickens. They are similar in the way they cause itchy bumps. Both fowl pox and chicken pox have to run their course, and both can be treated with some calamine lotion on the bumps to help the itching.

If you see your birds acting sick, point it out to an adult. That is the best thing you can do to help your chickens. Keep feeding them good, dry food and fresh, clean water; and you are the best "chicken tender" you can be.

FLOCK UNIVERSITY

COLOR THE SHIP

COCKA DOODLE DOOOOOOOOOO

Connect the dots

Things Traded on the Silk Road

```
G A E L K S E H S L E N O A K E S I L K W R O C H S L R H G M A
S K E O G D O D K G R N V L J D F K F N D I G A E P I D
Q R Y C G A R D K O Y H V X A O P W J V Y D G Q M V U K S
J A D E G S S T I G J L A V W O P F T H S L D N Q U C I B
G H A L I C E G H S J U I A C E D O P Q M V X Y H B E D S O
L O L A D H S M E A J D V U J M S I G E V F U R S V B E Q P L
G J U M S F T U W X V N O M U N S T F V J E P L A N V U F H I
I D L J E H D B Y F H S K C U P A P E R G N V I K L S T V G L
N L A G J T D I W Q M C O X B U G D K E L G B I H T I M C G
G W I G B D S U V M A P L V U T N U T M E G N S K I Y Q X O F
E G J D K V Y T B A P E M C Y G H J K L W S U V S X U L I H
R S J A S K V I F J D L W N X I L P A C R F B Q I F T H V
W H G I S M V B R G S J F O L A P E B C Y H E K A O I V D U
B C K S U G J E O Q P X U B N E M I K S F G H E V A O V B T
M S I K L P W F C U A T H B M K E L I X D G H Q P O Z M V U
B L A C K P E P P E R X S H U L A B V I L P S F B H J A Z I
N S H R J U A O P E C V I B C I N N A M O N F H T I C H E P
C N A L Y E I P Q X C U K A S L N I F G E J Q O Z P B N E
A M V U T G X E O L A B U T I D M P G L A S S W A R E C M W
```

Word Bank:

Horses, Silk, Paper, Jade, Ginger, Furs, Nutmeg, Cinnamon, Glassware, black pepper

Biomes

Biomes are a way to group different habitats around the world that share similar characteristics. They are defined by the plants and animals that live in each type of environment. Below are 4 types of biomes. You job is to draw a line from each picture to the matching definition. Then mark each pink box with the number of the chicken breed that would do best in that biome from the list below.

 #

Coniferous Forest Biome

This biome is defined by coniforous trees, which are trees with needles and cones. They don't drop needles in the winter. The temperature is colder than the temperate forest. The animals you find include things like bear, moose, and elk.

Rainforest Biome

This biome is defined by being wet. Heavy rainfall and tall tree canopies (top of the trees make a "ceiling" called a canopy) create a humid environment. Plant life is often very green and the animals found in this biome need the trees to survive. The animals and plants here are often colorful and sometimes poisonous like poison dart frogs.

 #

Temperate Diciduous Forest Biome

This biome has trees that lose their leaves in winter and are defined by 4-6 months of frost free weather. These forests can have pine trees too. Temperatures are moderate. Plants like berries and trees are abundant here. Some animals you will find are deer, squirrels, and many types of birds.

 #

Desert Biome

Very dry and usualy hot, but more importantly dry. The plants and animals that live in desert biomes have adapted to the extreme enviroment they live in. Plant life is sparse but adapted to conserve water. Cactus and succulants are some plants you can find. Animals found in this biome range from snakes to camels.

 #

#1 Dominique   #2 Egyptian Fayoumi   #3 Red Junglefowl   #4 Appenzeller Spitzhauben

Chicken eggs come in different colors. Even blue and green! Lets do some counting.

1.

There are ____ brown eggs.
There are ____ blue eggs.
There are ____ eggs all together.

In math, this looks like...

___ + ___ = ____

2.

There are ____ brown eggs.
There are ____ white eggs.
There is ____ blue egg.

In math, this looks like...

___ + ___ + ___ = ____

3.

There are ____ white eggs.
There are ____ other eggs.

In Math, this looks like...

___ + ___ = ____

4.

There are ____ white eggs.
There is ____ blue egg.
There is ____ dark brown egg.
There is ____ light brown egg.

In Math, this looks like...

___ + ___ + ___ + ___ = ____

USE THE PICTURE FROM CHAPTER 1 TO COLOR THE EGYPTIAN INCUBATOR BELOW. GET CREATIVE AND ADD EGGS, CHICKS, DUNG AND MAYBE EVEN A HUMAN WORKING ON THEM!

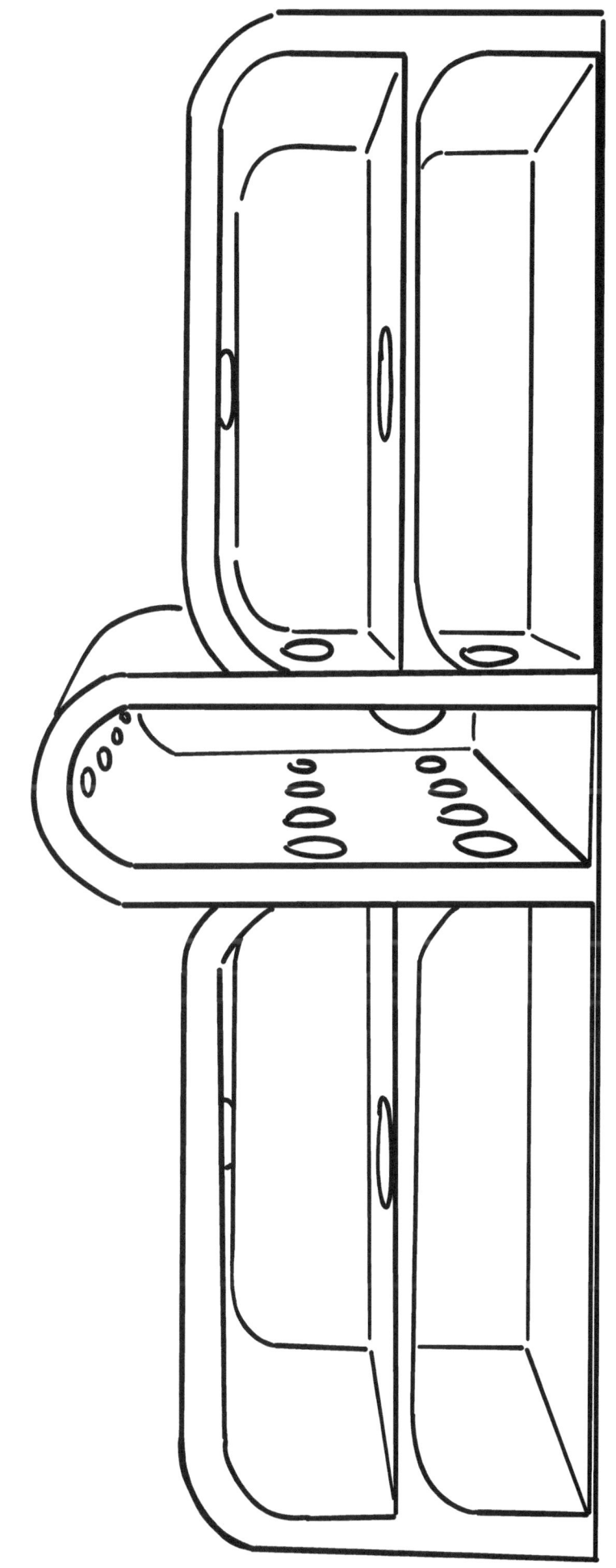

Grow your own rice!

Things you will need:

5 gallon bucket
Brown rice
Bowl
Paper towel
Potting soil
and
Permission from an adult

Step 1. Take brown rice (must be brown) and soak it in warm water in a bowl overnight.

Step 2. Place the rice inside two sheets of damp paper towls and place all that inside a plastic bag to keep in the moisture. This step should take two or three days. It's time to take them out when the sprouts are 2 inches long.

A US Quarter is about an inch in width.

Step 3. Take a 5 gallon bucket or something of that size. If it has holes in it, you have to close them up. Rice thrives in flooding environments.

Step 4. Put 6 inches of soil in the bottom of the bucket.

Step 5. Fill bucket with water until water is 2 inches above the soil.

Step 6. Take the sprouts from the paper towel and gently place them on the soil about 1 inch apart from one another. You may have to push them into the soil, but do not bury them. You should still see the rice.

Step 7. Place the bucket in a sunny location. The plants need the warmth and the light for about 7 hours a day.

Step 8. Very important step. Keep the water filled in the bucket 2 inches above the soil until the plants are about 6 inches tall, then make the water 4 inches above the soil. Keep the water topped off. This replicates the rainy season.

Step 9. When the plants are 8 inches tall, stop filling the bucket with water and let it all dry out as the plants continue to grow. This simulates a dry season.

The rice will be ready to be harvested once the plant turns brown. Make sure to properly harvest the rice if you plan on eating it. It needs to be fully dried to be safe to eat. By cutting the stalks of the plant and hanging them upside down to dry on a line, you can get a lot of the moisture out. You also can use a dehydrator. Once they are competely dry you can seperate the hulls from the edible rice by rubbing in your hands. the hulls and rice will seperate and you can then eat the rice the way you normally would.

Roman Numerals

Roman numerals are how Romans wrote numbers. We still use them. They used letters like "V", "I", and "X" to show numbers. How this works is each letter is worth a value. "X" is 10, "V" is 5, and "I" is 1 and where they are in the sequence determines if they are added or subtracted from the other values. For example, "IV" is 4. The "I" is placed in front of the "V" so you take 1 away from 5. However, "VI" is 6. The "I" (worth 1) is after the "V" (worth 5) so you add 1 to 5 and that is 6. Use this new knowledge to write the numbers displayed in Roman numerals below as you would in our common way of showing mathematics.

I = 1

V = 5

X = 10

III =

IV =

XX =

XII =

XV =

VI =

IX =

II =

XIV =

VIII =

XVII =

Label the Life Cycle

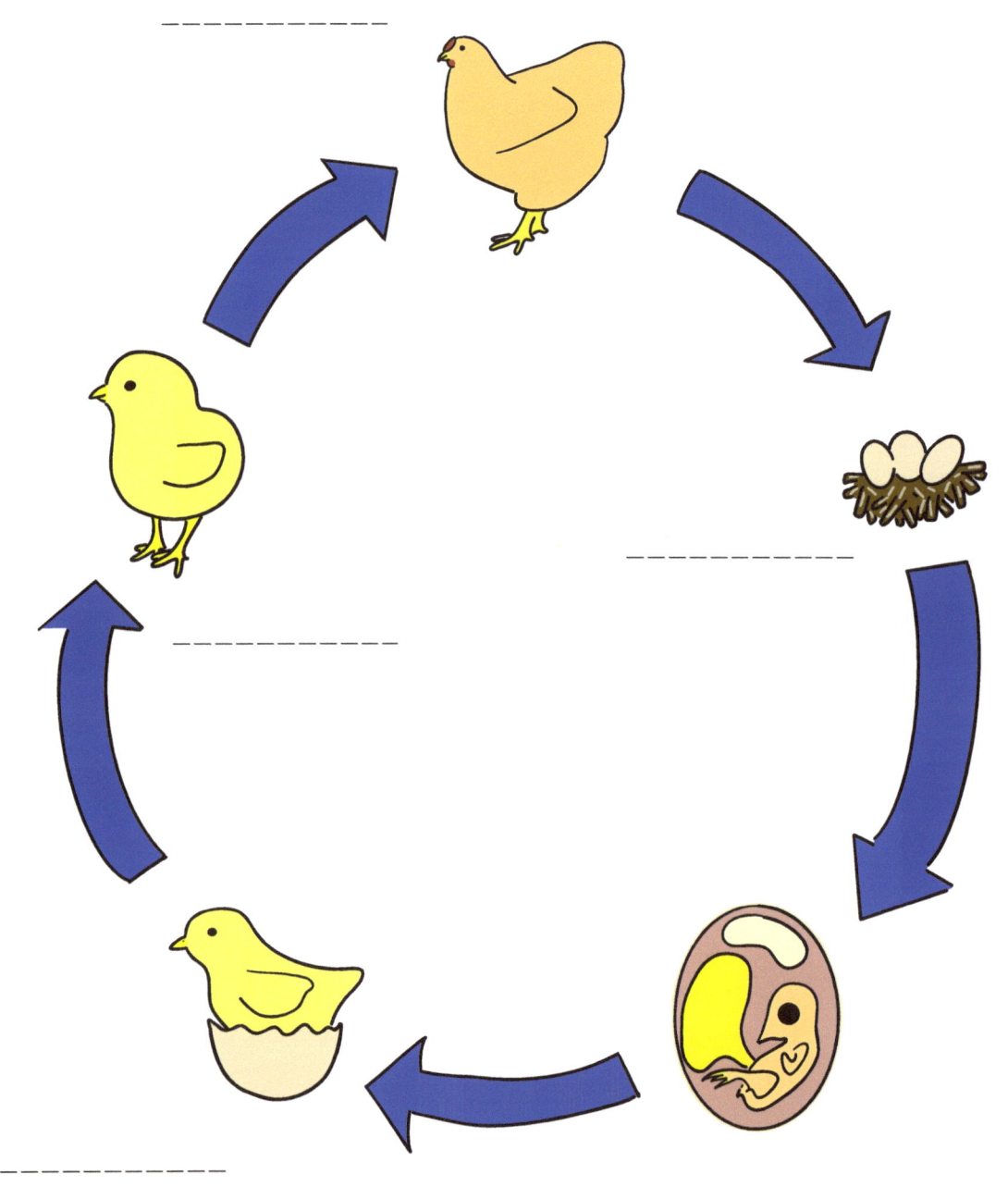

Candling

Let's pretend you are candling some eggs you had been incubating for about a week now. Count them and see how your hatch is going.

How many eggs did you originally start with? _____

How many should you take out? _____

How many are still doing good and growing? _____

Color the hen with chicks

Reading Thermometers

Thermometers have numbers up the side and fluid inside that will rise up to the temperature of the environment you are trying to measure. An incubator should be at 99.5 F. Which thermometer is correct for hatching chickens?

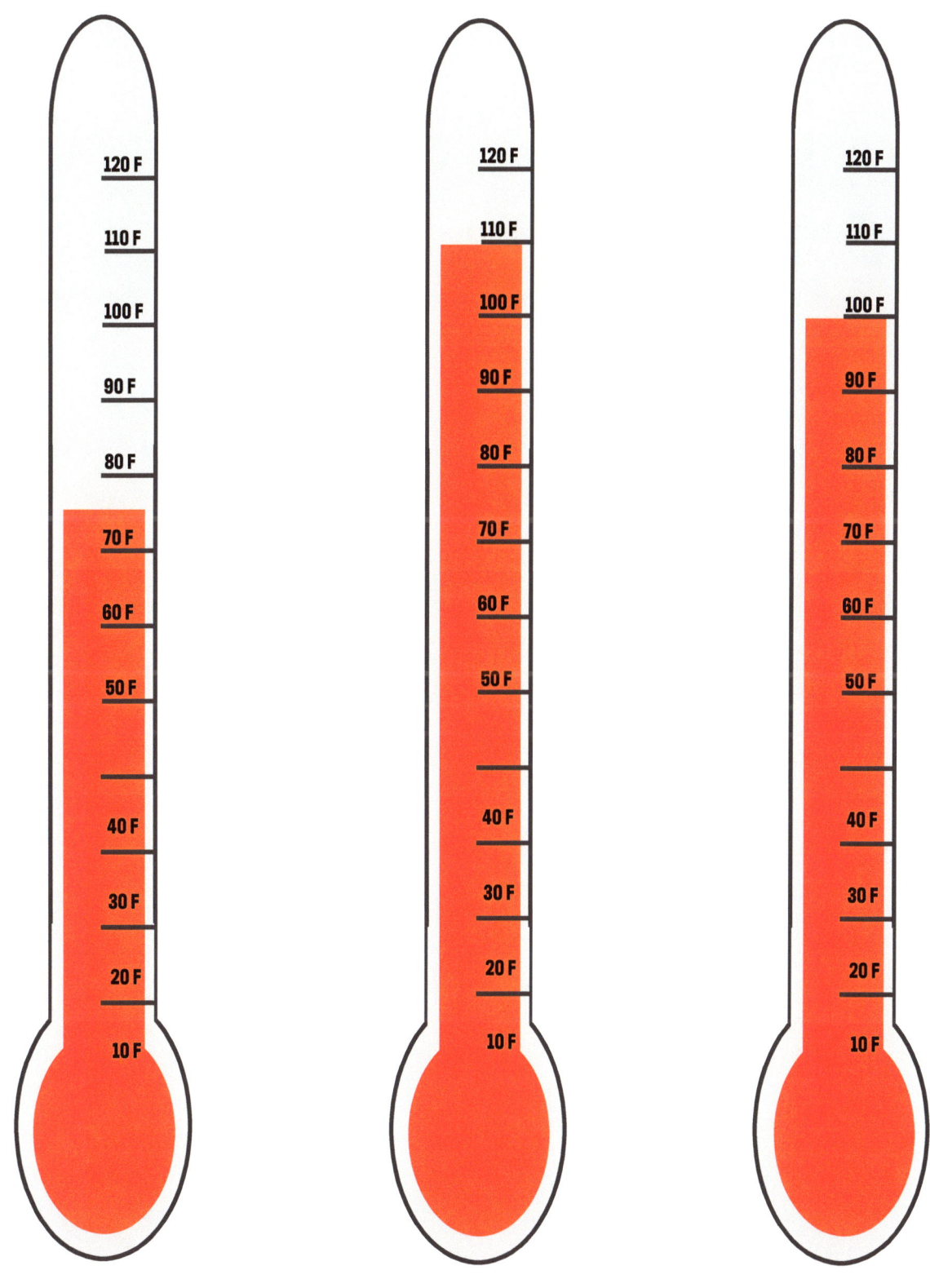

Match the bird to the feed

Crossword Puzzle: Chicken Breeds

Down:

1. An Egyptian breed that lays white eggs and does well in extreme heat
2. A true bantam named after the country it comes from. It has short legs.
4. A breed from Holland that lays dark brown eggs
6. The only breed developed by a woman
7. An American breed thought to be the first. It comes in one color called "cuckoo" which looks like black and white stripes.
10. An Asian breed with feathered feet and loved by Queen Victoria

Across:

3. A breed bred from Japanese breed called Minohiki
5. A Swiss breed with a unique comb
8. A English breed named after where it comes from, it has a color called "Coronation"
9. An American breed named after a Native American tribe
11. An ancient breed from the Roman Empire used for meat and eggs
13. An American breed named after the state it comes from and lays jumbo brown eggs

Egg Muffins

Ingredients:

12 eggs
1/4 cup whole milk
Salt and pepper to taste (start at 1/4 teaspoon)
1/2 lb sausage (or chicken, turkey, or pork)
1/2 cup shredded cheese

optional:

broccoli, jalapenos, tomatoes, bacon, chives, mushrooms, hashbrown potatoes, spinach, bell peppers

Prep:
Preheat oven to 350F
Brown and season meat of choice as you would like
Mix eggs with milk and salt and pepper in bowl with wisk
Butter or cooking spray muffin pan to keep eggs from sticking
Lay in spoonful of cooked meat
Pour egg mix into muffin mold, leaving room for rising
Top with cheese and additional toppings from optional list above

Cooking:
Bake tray at 350F for 20-25 minutes
Muffins are done when fork can be inserted in muffin and come out clean

Top with choice of sour cream, ketchup, hot sauce, scallions, or whatever you like!

Leftovers:
Leftovers are safe in air tight containers in the fridge for up to 5 days. Reheat in microwave for 1-2 minutes or oven in loose tin foil at 350F for until heated through.

Catching chickens

Catching chickens is quite the challenge for people. Chickens run at about 9 miles per hour and humans run at about 8 miles per hour. On top of them being faster, they can see 300 degrees around (a full circle is 360 degrees) and we humans can only see about 200 degrees. Sounds like it's impossible to catch a chicken, huh?

One trick I use (it also keeps me from ever having roosters that want to flog me) is what I coined a "poultry pace" when I walk. I always keep this pace when in the coop or around the chickens. It is a nice slow walk that is deliberate and mindful. Chickens tend to react to speed with speed. The rooster will perceive you as a threat and roosters attack threats. This poultry pace keeps everyone, including me, calm.

Use your poultry pace to slowly and quietly guide the chicken you want into a corner or wall. Line yourself up to be in the 60 degrees they don't see (basically directly behind them). Slowly (keeping the poultry pace) start to get low and have a hand ready to be where the chicken is going. Remember, you are not faster, you have to be smarter. Have a hand ready to be where the chicken will likely go and ready to stop the chicken with that hand on the chicken's chest. Your other hand will be able to be a second late and placed on the chicken's back. Now you have caught the chicken!

Catching chickens can be necessary when they need to be looked over for their health or maybe bathed before a show or just moved from one place to another. If you are needing more than one chicken, it may be best to wait until dark or get up early. Chickens are slower in the dark.

Join our Chicken Catching Team at Flock University

Check out flockuniversity.org for instructions

FLOCK UNIVERSITY

Follow us on socials!

 facebook.com/schoolofflock

 @flockuniversity

 @FlockUni

and flockuniversity.org